MIX
Papier aus verantwortungsvollen Quellen
Paper from responsible sources
FSC® C105338

Dr. Christo Ananth

Cardiac Patients Monitoring at a Distance

Anchor Academic
Publishing

Ananth, Christo: Cardiac Patients Monitoring at a Distance, Hamburg, Anchor Academic Publishing 2017

Buch-ISBN: 978-3-96067-195-4
PDF-eBook-ISBN: 978-3-96067-695-9
Druck/Herstellung: Anchor Academic Publishing, Hamburg, 2017

Bibliografische Information der Deutschen Nationalbibliothek:
Die Deutsche Nationalbibliothek verzeichnet diese Publikation in der Deutschen Nationalbibliografie; detaillierte bibliografische Daten sind im Internet über http://dnb.d-nb.de abrufbar.

Bibliographical Information of the German National Library:
The German National Library lists this publication in the German National Bibliography. Detailed bibliographic data can be found at: http://dnb.d-nb.de

All rights reserved. This publication may not be reproduced, stored in a retrieval system or transmitted, in any form or by any means, electronic, mechanical, photocopying, recording or otherwise, without the prior permission of the publishers.

Das Werk einschließlich aller seiner Teile ist urheberrechtlich geschützt. Jede Verwertung außerhalb der Grenzen des Urheberrechtsgesetzes ist ohne Zustimmung des Verlages unzulässig und strafbar. Dies gilt insbesondere für Vervielfältigungen, Übersetzungen, Mikroverfilmungen und die Einspeicherung und Bearbeitung in elektronischen Systemen.

Die Wiedergabe von Gebrauchsnamen, Handelsnamen, Warenbezeichnungen usw. in diesem Werk berechtigt auch ohne besondere Kennzeichnung nicht zu der Annahme, dass solche Namen im Sinne der Warenzeichen- und Markenschutz-Gesetzgebung als frei zu betrachten wären und daher von jedermann benutzt werden dürften.

Die Informationen in diesem Werk wurden mit Sorgfalt erarbeitet. Dennoch können Fehler nicht vollständig ausgeschlossen werden und die Diplomica Verlag GmbH, die Autoren oder Übersetzer übernehmen keine juristische Verantwortung oder irgendeine Haftung für evtl. verbliebene fehlerhafte Angaben und deren Folgen.

Alle Rechte vorbehalten

© Anchor Academic Publishing, Imprint der Diplomica Verlag GmbH
Hermannstal 119k, 22119 Hamburg
http://www.diplomica-verlag.de, Hamburg 2017
Printed in Germany

ABSTRACT

Nowadays heart attack patients are increasing day by day."Though it is tough to save the heart attack patients, we can increase the statistics of Saving the life of that patients & the life of others whom they are responsible for. The main design of a project is, in order to track the heart attack patients suffered any attacks during driving and sent them a medical need & to stop the vehicle which he or she is riding to ensure that the persons along them are safe from accident. An eye blinking sensor used to sense the blinking of the eye. And a spO_2 sensor to check the pulse rate of the patient. Both are connected to micro controller.

If eye blinking gets stopped then the signal is sent to the controller to make an alarm through the buffer. If spO_2 sensor senses a variation in pulse or low oxygen content in Blood it may results in heart failure then the controller stops the motor of the vehicle. Then Tarang F4 transmitter is used to sent the vehicle & the mobile number of the patient to a nearest medical station within 25 km for medical aid. The pulse rate monitored via LCD .The Tarang F4 receiver receives the signal and passes through controller, the number gets displayed in the LCD screen and an alarm is produced through a buzzer as soon the signal is received.

Five topics are discussed in this project : Detecting the patient BPM and the Eye blinking status; Transmitting via Tarang F4 in case of abnormalities in patient; The patient status is displayed and indicated by Buzzer; The Hospital Unit receives the patient's mobile and the car number; The communication between the vehicle and the Hospital through Tarang F4.

TABLE OF CONTENTS

CHAPTER NO.	TITLE	PAGE NO.
	ABSTRACT	i
	LIST OF TABLES	vi
	LIST OF FIGURES	vii
	LIST OF ABBREVIATIONS	ix
1.	INTRODUCTION	1
	1.1 REALTIME MONITORING OF CARDIAC PATIENTS	1
	1.2 REALTIME MONITORING OF CARDIAC PATIENTSUSING EMBEDDED SYSTEMS	1
	1.3 IMPORTANT STAGES IN EMBEDDED SYSTEM DESIGN	2
	1.3.1 Requirement Analysis	2
	1.3.2 System Specification	2
	1.3.3 Architecture Design	2
	1.3.4 Design of Software and Hardware Components	3
	1.3.5 System Integration	3
	1.4 BLOCK DIAGRAM OF THE EMBEDDED SYSTEM	3
	1.4.1 Sensor	3
	1.4.2 Signal Conditioning Unit	4
	1.4.3 Processor	4
	1.4.4 Receiver and display section	4
	1.4.5 Operator console	4
	1.5 CONCLUSION	4

2.		**LITERATURE SURVEY**	5
	2.1	INTRODUCTION	5
	2.2	EXISTING SYSTEM	14
		2.2.1 Disadvantages	14
	2.3	CONCLUSION	14
3.		**PROPOSED SYSTEM**	15
	3.1	INTRODUCTION	15
	3.2	BLOCK DIAGRAM	16
		3.2.1 Block diagram of Car Unit	16
		3.2.2 Block Diagram of Hospital Unit	17
	3.3	BLOCK DIAGRAM DESCRIPTION	18
		3.3.1 Car Unit	18
		3.3.2 Hospital Unit	18
	3.4	CIRCUIT DIAGRAM	19
		3.4.1 Circuit Diagram of Car Unit	19
		3.4.2 Circuit Diagram of Hospital Unit	20
	3.5	HARDWARE REQUIREMENTS	21
	3.6	SOFTWARE REQUIREMENTS	21
	3.7	AT MEGA 8535 MICROCONTROLLER	21
		3.7.1 Pin Description	22
		3.7.2 ATMEGA 8535 Architecture	24
		3.7.3 Memory Organization	25
		3.7.4 SRAM Data Memory	26
		3.7.5 Special Function Registers (SFR)	26
		3.7.6 Program Counter	26
		3.7.7 Input/output ports (I/O Ports)	26
		3.7.8 Status Register	27

3.8	AT MEGA 8 MICROCONTROLLER	27
	3.8.1 Pin Description	28
	3.8.2 ATMEGA 8 Architecture	30
	3.8.3 Memory Organization	32
	3.8.4 Special Function Registers (SFR)	32
	3.8.5 Program Counter	32
	3.8.6 Input/output ports (I/O Ports)	32
3.9	TARANG	32
	3.9.1 Features	33
3.10	LCD	34
	3.10.1 LCD Technologies and types	34
	3.10.2 LCD characteristics	35
	3.10.3 Advantages	35
3.11	BUZZER	35
	3.11.1 Features	36
3.12	IR SENSOR	36
	3.12.1 Features	36
3.13	LED	37
	3.13.1 Advantages Of LED	38
3.14	POWER SUPPLY	38
3.15	VOLTAGE REGULATOR: (IC 7805)	39
	3.15.1 Features	39
	3.15.2 Description	39
3.16	RELAY	39
3.17	KEIL COMPILER	40
3.18	EMBEDDED C	43
3.19	CONCLUSION	43

4.	**RESULTS AND DISCUSSION**	**44**
	4.1 INTRODUCTION	44
	4.2 CAR UNIT	44
	4.2.1 Inference	44
	4.2.2 Eye Blinking IR Sensor	45
	4.2.3 Heart Beat IR Sensor	46
	4.2.4 TarangF4	47
	4.2.5 Working	48
	4.3 HOSPITAL UNIT	50
	4.3.1 Inference	50
	4.3.2 Buzzer	50
	4.3.3 LCD Display	51
	4.3.4 Working	51
5.	**CONCLUSION AND FUTURE ENHANCEMENT**	**52**
	5.1 CONCLUSION	52
	5.2 SCOPE FOR FUTURE ENHANCEMENT	52
	5.3 APPLICATIONS	52
	REFERENCES	**53**
	APPENDIX	**55**

LIST OF TABLES

TABLE NO.	TABLE NAME	PAGE NO.
3.1	Status Register	27

LIST OF FIGURES

FIG NO.	TITLE	PAGE NO.
1.1	Block Diagram of a Typical Embedded System	3
3.1	Block Diagram of Car unit	16
3.2	Block Diagram of Hospital unit	17
3.3	Circuit Diagram Of Car Unit	19
3.4	Circuit Diagram Of Hospital Unit	20
3.5	Pin Diagram Of ATMEGA 8535	22
3.6	AT MEGA 8535 Architecture	25
3.7	Pin Diagram of ATMEGA 8	28
3.8	AT MEGA 8 Architecture	31
3.9	Tarang	33
3.10	LCD	34
3.11	Structure Of LED	37
3.12	Circuit diagram of power supply	39
3.13	Relay	40
4.1	Image of Car Unit	44
4.2	IR Transmitter	45
4.3	IR Receiver	45
4.4	LM 358 IC Connected With Eye Blinking Sensor Circuit	46

4.5	A LM358 Dual Op-Amp	46
4.6	Pulse Detection sensor	47
4.7	Tarang Module	47
4.8	Top View of Transmitter Car Unit	48
4.9	Front View of Transmitter Car Unit	49
4.10	Image Of Hospital Unit	50
4.11	Buzzer	51
4.12	LCD	51

LIST OF ABBREVIATIONS

A/D	Analog to Digital
ALU	Arithmetic Logic Unit
BPM	Blood Pressure Monitor
CMOS	Complementary Metal Oxide Semiconductor
ECG	Electrocardiogram
EEG	Electroencephalogram
GPS	Global Positioning System
GSM	Global System For Mobile
HRV	Heart Rate Variability
IR	Infrared
ISM	Industrial Scientific Medical
LCD	Liquid Crystal Display
LED	Light Emitting Diode
PC	Program Counter
PSU	Power Supply Unit
PUCK	Potential Of Unbalanced Complex Kinetics
RAM	Random Access Memory
SFR	Special Function Registers
WECG	Wide Fidelity
Wi-Fi	Wireless Electrocardiograph

CHAPTER 1

INTRODUCTION

1.1 REALTIME MONITORING OF CARDIAC PATIENTS

Real time monitoring of cardiac patients are used to protect the heart patients, including alert systems and the vehicle onboard controllers with components that communicate with each other. Realtime monitoring of cardiac patients are typical safety systems since they prevent the death of heart patients and ensure the safety operations in general.

1.2 REALTIME MONITORING OF CARDIAC PATIENTS USING EMBEDDED SYSTEMS

Here we implent a cardiac patient monitoring system in which at the first level choose the Tarang F4 for the communication which has more coverage area than the zigbee, for the communication between the vehicle and Hospital. Hence here we design a system which can be used by the vehicle and the Hospital which can communicate each other.We design an automated communication between the patient to Hospital.We also design the automatic vehicle ignition OFF. As the patient suffers any attack or he/she is sleeping,the buzzer indicates an alarm and the patient's BPM, eye blinking ratio is dislayed in LCD as "EYE CLOSED" or NOT.The IR sensors monitors the both function. Meanwhile the patient's status is transmitted to the hospital. The mobile number and vehicle number is displayed on the hospital LCD. And is indicated by an alarm and RED LED. The Hospital members then tracks the patient.[1]

1.3 IMPORTANT STAGES IN EMBEDDED SYSTEM DESIGN

- Requirement analysis.
- System specification.
- Architecture design.
- Design of software and hardware components.
- System integration.

1.3.1 Requirement Analysis

- To gather informal description for the specified task.
- The sensors output are to be monitored in coordination with the GPS output in simultaneous manner.
- The basic requirements are analyzed the area specification in meters, the length of the power line fence, the distance between the control section and the fence section, the sensor network alignment specification are noted.

1.3.2 System Specification

- The specification is more precise as it serves the requirements.
- The specification in this system is to analyze the sensor needed for the specific tasks and the GPS specifications.

1.3.3 Architecture Design

- Architectural descriptions must be designed to specify both functional and non-functional requirements
- The basic flow and the system block diagrammatic evaluation are made in this process.
- The control section must include two monitoring units. The fence section includes the sensor networks.
- The outputs of the fence section are given to the control section for processing.

1.3.4 Design of Software and Hardware Components

- The component design effort builds those components in conformance to the architecture and specification. The components will in general include both the hardware and the software specifications.
- The processor needed for the sensor networks as well as the monitoring system specifications are provided in this section of analysis.
- The two section outputs are defined with the requirement and are cross evaluated.

1.3.5 System Integration

- After the components are built to the satisfaction they are put together by using buses and the sensor network is placed at the specified location.
- The debugging process is carried out here, the system various observations are made and the specified localization output is noted.
- The system is pre-evaluated.

1.4 BLOCK DIAGRAM OF THE EMBEDDED SYSTEM

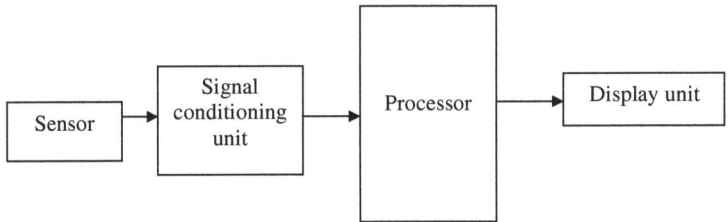

Fig 1.1 Block Diagram of a Typical Embedded System

1.4.1 Sensor

A Sensor is an important part of the embedded system its main function is to gather the information from its specified environment and proceeds to the processing units for evaluation.

1.4.2 Signal Conditioning Unit

The sensor output is provided to the signal conditioning unit, the function of signal conditioning unit is to match the obtained sensor output with the processor requirements. There are many components in the signal conditioning section. The basic components are the amplifier, analog to digital converter.

1.4.3 Processor

A Processor is a programmable device that processes information by manipulating sensor outputs according to logical rules. Arithmetic processing is also done in this section. The processor is called as the heart of the system. In this project we are proposing microcontroller ATMEL AT89S52 as the processor.

1.4.4 Receiver and display section

The display section is used as a user interface with the processors output. the receiving section includes the sets of buses and display units which contribute to the output of the processor being evaluated.

1.4.5 Operator console

The operator is m capable of checking for any resulting errors and for the entry of requisite data. The operator console is a desktop application that operators use to communicate with web visitors. It runs on Window and Linux.

1.5 CONCLUSION

Thus the introduction of our project and introduction to embedded system is discussed detail in this chapter.

CHAPTER 2

LITERATURE SURVEY

2.1 INTRODUCTION

The following papers have been studied for their working, advantages and limitations.

[1] Design And Development of a Wireless Remote Point- of-Care Patient Monitoring System, Whitchurch, A.K.; Abraham, J.K.; vol., no., pp.163,166.
Inference:

Whitchurch, A.K proposed a system describes the point of care patient monitoring system. Remote patient monitoring is an alternative to regular home check-ups of patients with certain special medical conditions or the elderly who are unable to regularly visit a healthcare facility. This technology reduces the number of home visits which are now only required when special attention is needed. This paper presents the design and development of a remote point-of-care patient monitoring system which allows the patient to be monitored remotely while remaining in the comfort of their home.The system described here allows wireless data acquisition from eight patient-worn sensors. The number and type of sensors are configurable according to the subject's specific condition. The system uses the standard Bluetooth technology for communication with a home based monitor which in turn relays this data to the remote healthcare facility using the internet. This data can be used for real-time evaluation of the patient's conditions as well as data logging for later analysis. Since this is a configurable system, a few selected sensors are connected to demonstrate the concept of remote patient monitoring; these include Electrocardiogram (ECG), Electroencephalogram (EEG), Airflow, respiration, patient movement and body temperature. The results obtained from these tests are also presented in this paper.

Advantages of this system

- An alternative to regular home check-ups of patients with certain special medical conditions or the elderly who are unable to regularly visit a healthcare facility.

Limitations

- This method is suitable for only the patient is not in motion.

[2] Wireless ECG Monitoring and Alarm System Using Zigbee, Apostu, O.; Hagiu, B.; vol., no., pp.1,4.

Inference:

Apostu, O. proposed a system describes the ECG Monitoring using Zigbee. This paper presents the development of a system for wireless ECG monitoring and alarm using ZigBee. The system is intended for home use by patients that are not in a critical condition but need to be constant or periodically monitored by clinicians or family. Patient monitoring is the cornerstone of proper medical care. It provides clinicians the much needed information about a person's current health status, so that they can act accordingly if anything goes wrong. Nowadays, complex patient monitoring systems offer the possibility of continuously monitoring a multitude of biological signals, analyze them, interpret them and take the appropriate action; or alert clinicians if necessary. The usual shortcomings of most of these systems reside in affecting patient mobility and home comfort. A patient would need to be sitting on a bed wired to these devices in order for his vital signs to be monitored. This system measures, records and presents in real-time the electrical activity of the heart while preserving comfort of the patient. The device is built as a low-power, small-sized, low-cost solution suitable for monitoring elderly people at home or in a nursing facility without interfering with the daily activity of a patient. It should give sufficient information in real time, and make it available remotely. The intention is not to achieve perfect clinical accuracy but the device is able to detect anomalies in the measured data and it also has alerting features. Authorized observers (clinicians or family) can monitor at any moment the state of the patient through the internet.

Advantages of this system
- It is intended for home use by patients that are not in a critical condition but need to be constant or periodically monitored by clinicians or family.
- Low-power, small-sized, low-cost solution suitable for monitoring elderly people at home.

Limitations
- This method is not applicable for the patient in movement.
- No alarm or any alert indication.

[3] Weight and Activity With Blood Pressure Monitoring System For Heart Failure Patients, Myung-kyung Suh; Evangelista, L.S.; Chen, vol., no., pp.1,6, 14-17 June 2010

Inference:

Myung-kyung Suh proposed a system describes the Blood Pressure Monitoring System For Heart Failure Patients. Heart failure is a leading cause of death in the United States, with around 5 million Americans currently suffering from congestive heart failure. The WANDA B. wireless health technology leverages sensor technology and wireless communication to monitor heart failure patient activity and to provide tailored guidance. Patients who have cardiovascular system disorders can measure their weight, blood pressure, activity levels, and other vital signs in a real-time automated fashion. The system was developed in conjunction with the UCLA

Nursing School and the UCLA Wireless Health Institute for use on actual patients. It is currently in use with real patients in a clinical trial.

Advantages of this system

- The patient's weight, blood pressure, activity levels, and other vital signs are measured in a real-time automated fashion.

- Immediate action to patient's status.

Limitations

- This method is suitable for the patient is Hospitalized.

[4] A Mobile Care System With Alert Mechanism ,Ren-Guey Lee; Yih-Chien Chen; vol.11, no.5, pp.507,517, Sept. 2007

Inference:

Ren-Guey Lee describes the physiological parameters of the patient are constantly monitored. Hypertension and arrhythmia are chronic diseases, which can be effectively prevented and controlled only if the physiological parameters of the patient are constantly monitored, along with the full support of the health education and professional medical care. In this paper, a role-based intelligent mobile care system with alert mechanism in chronic care system include patients, physicians, nurses, and health care providers. Each of the roles represents a person that uses a mobile device such as a mobile phone to communicate with the server setup in the care center such that he or she can go around without restrictions. For commercial mobile phones with Bluetooth communication capability attached to chronic patients, we have developed physiological signal recognition algorithms that were implemented. It is thus possible to integrate several front-end mobile care devices with Bluetooth communication capability to extract patients' various physiological parameters [such as blood pressure, pulse, saturation of haemoglobin (SpO_2), and electrocardiogram (ECG)], to monitor multiple physiological signals without space limit, and to upload important or abnormal physiological information to healthcare center for storage and analysis or transmit the information to physicians and healthcare providers for further processing. Thus, the physiological signal extraction devices only have to deal with signal extraction and wireless transmission.

Advantages of this system

- Physiological parameters of the patient are constantly monitored and they are periodically communicated to the health care providers.

Limitations

- This method is not suitable for the patient is driving.

[5] GSM Based Intelligent Wireless Mobile Patient Monitoring System Using Zigbee Communication, Kareti, Suresh Babu,ICEEE, 9 Sept 2012.

Inference:

    Kareti proposed a system the ever-growing age median among travelers, a health monitoring application is becoming more of a necessity in large capacity aircraft environments, providing safety to passengers with actual or chronic risks, and reducing risk and cost for long-range aircraft operations.

    Considering the technological advancements in embedded sensor devices a portable medical monitoring enclosure has been developed to provide with the flexibility of low cost and high accuracy measurement equipment in avionic environments. several types of health monitoring sensor modules can be integrated in a compact portable enclosure - such as Electrocardiogram, pulse rate, blood pressure, oximetry, temperature. In addition, a control board was designed and implemented with the purpose of interfacing and processing the data arriving from the sensor modules, and their transfer to a standard RS232 interface.

Advantages of this system
- Health monitoring application is mainly proposed to provide alerts for medical health monitoring staff for the patients when needed.
- It can be taken by patient and keep the patient movement intact because it is miniature and portable.

Limitations
- The GSM Signal fails at remote areas.
- ZigBee Coverage area is small.

[6] Development of An Online Database System For Remote Monitoring Of Artificial Heart Patient, Choi, J.; Kang, W. Y.; Chung, J.; Park, vol., no., pp.59,61, 24-26 April 2003

Inference:

Choi,J. proposed a system in which an online database system for remote monitoring of artificial heart patient has been developed. An online database system for remote monitoring of artificial heart patient has been developed. It is important for the patient with artificial heart implant to be discharged from the hospital after proper stabilization period for better and fast recovery and quality of life. The patients with artificial heart or mechanical circulatory support device are currently discharged after 1-2 months of hospitalization period. But, reliable and practical continuous remote monitoring systems for this kind of patients with life support device are still under study.

The authors have developed a database system for this purpose that consists of a portable monitoring terminal, a database for continuous recording of patient and device status, and a web-based data access system with which the clinicians can access the real-time registering of patient data and past history data as well. The system has been tested with data generation emulators installed on remote sites for simulation study and in 2 cases of actual animal experiment conducted at remote facilities. The system showed acceptable functionality and reliability. The results are expected to be a good base for practical use of the system. A machine intelligence based automatic data analysis module is to be included for better practicality.

Advantages of this system

- The database system for the continuous patient monitoring records.

Limitations

- This method is suitable for only the artificial heart patient.

[7] In-Home Wireless Monitoring of Physiological Data For Heart Failure Patients ,Mendoza, G. G.; Tran, vol.3, no., pp.1849,1850 vol.3, 23-26 Oct. 2002

Inference:

Mendoza, G.G. proposed a system in which the physiological data of the heart failure patient at home is monitored. Transducers for measurement of electrocardiogram (ECG), heart rate variability (HRV), acoustical data are embedded into patient clothing for unobtrusive monitoring for early, sensitive detection of changes in physiologic status. Sampling rate for this system is 1 kHz per channel. Signal conditioning is performed in hardware by the patient wearable system, after which information is wirelessly transmitted to a central server located elsewhere in the home for signal processing, data storage, and data trending. The dynamic frequency ranges for the ECG and heart sounds (HS) are 0.05-160 Hz and 35-1350 Hz, respectively. The range-of-operation for the current patient-wearable physiologic data capture design is 100±10 feet with direct line-of-sight to the home server station. Weight measurements are obtained directly by the in-home medical server using a digital scale. Physiologic information (ECG, HRV, HS, and weight) are dynamically analyzed using a combination o the LabVIEW (National Instruments, Inc.; Austin, TX) and MATLAB (MathWorks, Inc.; Inc; Natick, MA) software strategies. Software-based algorithms detect out-of-normal or alarm conditions for HR and weight as defined by the health care provider, information critical for HF patients. Health care professionals can remotely access vital data for improved management of heart failure.

Advantages of this system

- The electrocardiogram (ECG), heart rate variability (HRV), acoustical data are embedded into patient clothing for unobtrusive monitoring for early, sensitive detection of changes in physiologic status.

Limitations

- This paper is not suitable for the mobile patient's and it does not deals with the alert system.

[8] A 2.5 Ghz Wireless ECG System For Remotely Monitoring Heart Pulses, Palantei, E.; Baharuddin, vol., no., pp.1,2, 8-14 July 2012

Inference:

Palantei, E. proposed a system in which the patient's heart pulses is monitored remotely. A wireless electrocardiograph technology (WECG) which works on 2.5 GHz frequency band has been designed, implemented and evaluated in the actual environments both indoor and outdoor. WECG system has the benefit for improving the quality of the health care services especially for monitoring and evaluating people heart pulses record during the treatments. These are also practically useful to apply on a number medical cases, for instance, the heart disease treatment, early medical treatment at the disaster areas (e.g. earthquake, tsunami, the traffic accident on the road), other remote medical monitoring in health care centers, or intensive care units at the hospital. The WECG technology allows the medical authorities such as doctor and nurses capable to monitor the patients flexibly and immediately from a remote location. The read range testing of wireless ECG system to perform on monitoring the heart pulses was carried out. At indoor environments it can read up to the distance of more than 50 m. When operating at LOS outdoor environment the master unit and local ECG sensor unit can communicate for the distance longer than 250 m.

Advantages of this is system
- Patient's heart pulses is monitored remotely.
- The WECG technology allows the medical authorities such as doctor and nurses capable to monitor the patients flexibly and immediately from a remote location.

Limitations
- This method is suitable for patient under medical care.

[9] Cardiac Arrhythmia Detection Using Combination of Heart Rate Variability Analyses And PUCK Analysis, Mahananto, F.; Igasaki, T.; vol., no., pp.1696,1699, 3-7 July 2013

Inference:

Mahananto, F. proposed a system in which the Heart Rate Variability Analyses is done. This paper presents cardiac arrhythmia detection using the combination of a heart rate variability (HRV)analysis and a "potential of unbalanced complex kinetics" (PUCK) analysis. Detection performance was improved by adding features extracted from the PUCK analysis. Initially, R-R interval data were extracted from the original electrocardiogram (ECG) recordings and were cut into small segments and marked as either normal or arrhythmia. HRV analyses then were conducted using the segmented R-R interval data, including a time-domain analysis, frequency-domain analysis, and nonlinear analysis. In addition to the HRV analysis, PUCK analysis, which has been implemented successfully in a foreign exchange market series to characterize change, was employed. A decision-tree algorithm was applied to all of the obtained features for classification. The proposed method was tested using the MIT-BIH arrhythmia database and had an overall classification accuracy of 91.73%. After combining features obtained from the PUCK analysis, the overall accuracy increased to 92.91%. Therefore, we suggest that the use of a PUCK analysis in conjunction with HRV analysis might improve performance accuracy for the detection of cardiac arrhythmia.

Advantages of this system
- Cardiac arrhythmia detection using the combination of a heart rate variability (HRV)analysis and a "potential of unbalanced complex kinetics" (PUCK) analysis.

Limitations
- This method is suitable for the Hospitalized patients.

2.2 EXISTING SYSTEM

During the past few decades, the patient is Hospitalized or remote monitored in home through the hospital server by connecting the heart rate variation identifying components to the patient's body. If the heart patient's situation is critical it's not indicated by an alarm or buzzer. Only the database system for the continuous patient monitoring records is maintained.

2.2.1 Disadvantages
- The patient is Hospitalized or treated at home.
- The patients status is critical it is only transmitted to the Hospital server. It is not indicated by a buzzer.
- Not in Realtime monitoring (driving or in motion).
- Can't able to predict the heart attack.

2.3 CONCLUSION

Thus the Literature Survey describes the papers related to the project, and the Existing System is also described.

CHAPTER 3

PROPOSED SYSTEM

3.1 INTRODUCTION

A model-based development and verification approach for the Monitoring of Heart patients, and communication at the time of Heart beat variation along with the buzzer indication. The IR Sensor detects the patient's BPM and the Eye Blinking status. The Tarang F4 is used for the communication between the vehicle and the Hospital. The LCD in the Car unit displays the BPM and the Eye Blinking status. The LCD in the Hospital unit displays the patient's vehicle and the mobile number.

This project is divided into 2 units- Car unit and Hospital unit. The car unit is placed inside the car. It consists of Heart Beat sensor, Eye Blinking sensor, Tarang F4 Transmitter,and the DC Motor all are incorporated with AT Mega 8535controller. The Hospital unit consists of Tarang F4,connected to the AT Mega 8 controller. Both the units have Relay and Liquid Crystal Display. The power supply is provided by the 9V Adapter. The LCD used here is a 16*2 Display.

3.2 BLOCK DIAGRAM

3.2.1 Block diagram of Car Unit

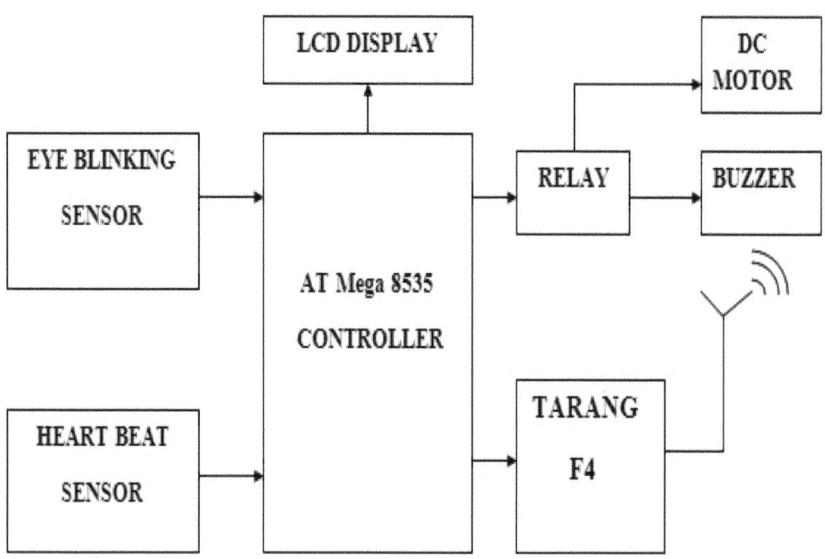

Fig 3.1 Block Diagram of Car unit

Fig.3.1 Shows The Transmitter Block Diagram Of the Circuit.(i.e)The Car Unit.

3.2.2 Block Diagram of Hospital Unit

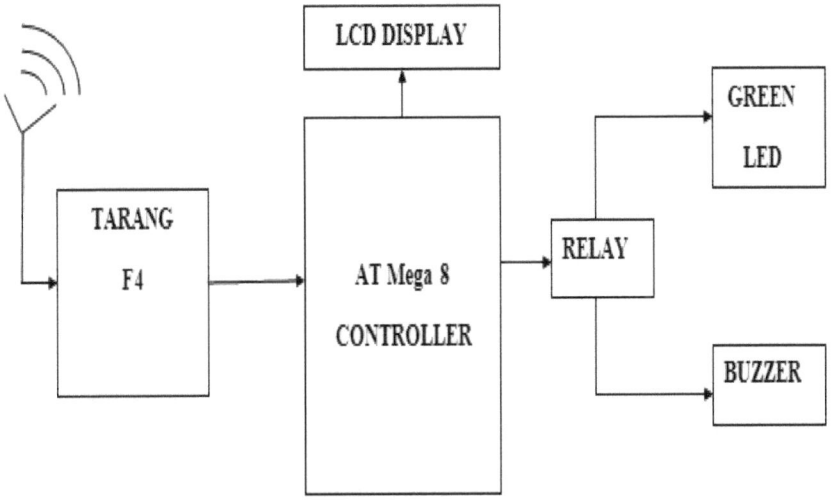

Fig 3.2 Block Diagram of Hospital unit

Fig.3.2 Shows The Receiver Block Diagram Of the Circuit (i.e) The Hospital Unit.

3.3 BLOCK DIAGRAM DESCRIPTION

3.3.1 Car Unit

The Car Unit consists of Microcontroller, Tarang F4, LCD Display, Relay, DC Motor, Eye Blinking (IR) Sensor, Heart Beat (IR) Sensor. . Here we use an eye blinking sensor to sense the eye blinking time, and a spo2 sensor to check the pulse rate of the patient. Both are connected to micro controller.

If eye blinking gets stopped then the signal is sent to the controller to make an alarm through the buffer. If spo2 sensor senses a variation in pulse or low oxygen content in Blood it may results in heart failure then the controller stops the motor of the vehicle. Then Tarang F4 transmitter is used to sent the vehicle no. & the mobile no. of the patient to a nearest medical station within 25 km for medical aid. The pulse rate monitored via LCD.

3.3.2 Hospital Unit

The Hospital Unit Consists of Microcontroller, Tarang F4, Relay, Green LED, LCD and the Buzzer. As soon as the signal received by the Tarang F4 it sends the data to the controller. The controller output is normally LOW, Therefore the Relay ON the Green LED. If the Controller output is High the Relay toggles to the Buzzer. The Controller displays the patient's Vehicle and the Mobile number .

The car unit have AT Mega 8535 Microcontroller and the Hospital Unit have AT Mega 8 Microcontroller.

3.4 CIRCUIT DIAGRAM

3.4.1 Circuit Diagram of Car Unit

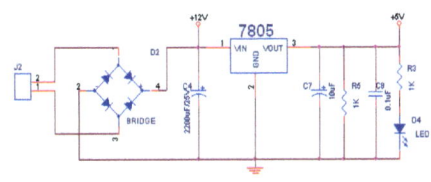

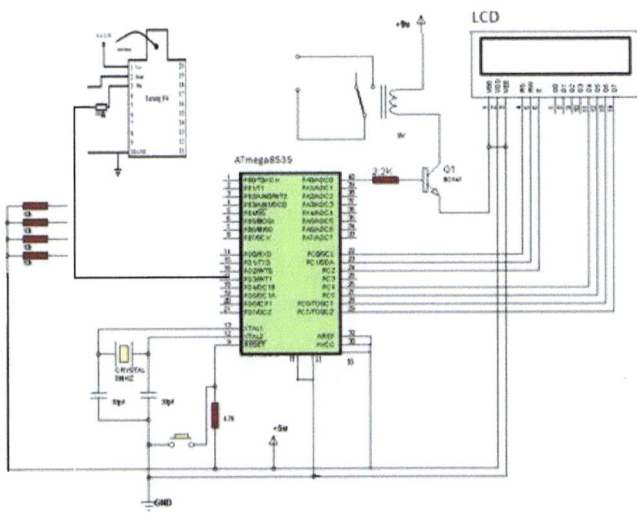

Fig 3.3 Circuit Diagram of Car Unit

Fig.3.3 Describes the Circuit Diagram of the Transmitter unit comprising of Microcontroller and LCD.

3.4.2 Circuit Diagram of Hospital Unit

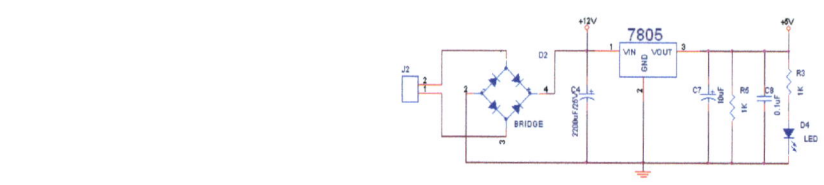

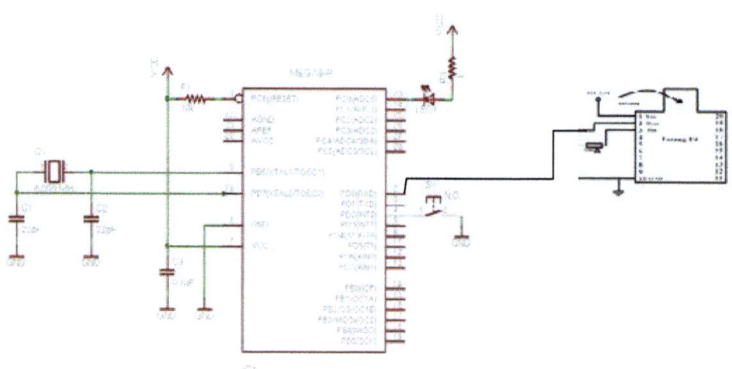

Fig 3.4 Circuit Diagram of Hospital Unit

Fig.3.4 Describes The Circuit Diagram Of The Receiver Unit.

3.5 HARDWARE REQUIREMENTS

- MICROCONTROLLER AT Mega 8535
- MICROCONTROLLER AT Mega 8
- EYE BLINKING IR SENSOR
- HEART BEAT IR SENSOR
- TARANG F4
- DC MOTOR
- BUZZER
- LCD DISPLAY
- IR SENSORS
- LED
- RELAY
- POWER SUPPLY

3.6 SOFTWARE REQUIREMENTS

- KEIL CROSS COMPILER
- EMBEDDED C

3.7 AT MEGA 8535 MICROCONTROLLER

The ATmega8535 is a low-power CMOS 8-bit microcontroller based on the AVR enhanced RISC architecture. By executing instructions in a single clock cycle, the ATmega8535 achieves throughputs approaching 1 MIPS per MHz allowing the system designer to optimize power consumption versus processing speed. The AVR core combines a rich instruction set with 32 general purpose working registers. All 32 registers are directly connected to the Arithmetic Logic Unit (ALU), allowing two independent registers to be accessed in one single instruction executed in one clock cycle. The resulting architecture is more code efficient while achieving throughputs up to ten times faster than conventional CISC microcontrollers.

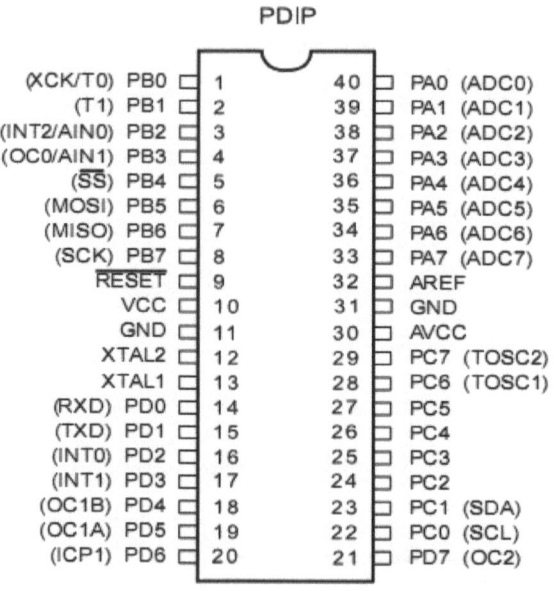

Fig 3.5 Pin Diagram Of ATMEGA 8535

3.7.1 Pin Description

VCC: Digital supply voltage.

GND: Ground.

Port A (PA7..PA0): Port A serves as the analog inputs to the A/D Converter. Port A also serves as an 8-bit bi-directional I/O port, if the A/D Converter is not used. Port pins can provide internal pull-up resistors (selected for each bit). The Port A output buffers have symmetrical drive characteristics with both high sink and source capability. When pins PA0 to PA7 are used as inputs and are externally pulled low, they will source current if the internal pull-up resistors are activated. The Port A pins are tri-stated when a reset condition becomes active, even if the clock is not running.

Port B (PB7..PB0): Port B is an 8-bit bi-directional I/O port with internal pull-up resistors (selected for each bit). The Port B output buffers have symmetrical drive characteristics with both high sink and source capability. As inputs, Port B pins that are externally pulled low will source current if the pull-up resistors are activated. The Port B pins are tri-stated when a reset condition becomes active, even if the clock is not running.

Port C (PC7..PC0): Port C is an 8-bit bi-directional I/O port with internal pull-up resistors (selected for each bit). The Port C output buffers have symmetrical drive characteristics with both high sink and source capability. As inputs, Port C pins that are externally pulled low will source current if the pull-up resistors are activated. The Port C pins are tri-stated when a reset condition becomes active, even if the clock is not running.

Port D (PD7..PD0): Port D is an 8-bit bi-directional I/O port with internal pull-up resistors (selected for each bit). The Port D output buffers have symmetrical drive characteristics with both high sink and source capability. As inputs, Port D pins that are externally pulled low will source current if the pull-up resistors are activated. The Port D pins are tri-stated when a reset condition becomes active, even if the clock is not running.

RESET : Reset input. A low level on this pin for longer than the minimum pulse length will generate a reset, even if the clock is not running. The minimum pulse length is given Shorter pulses are not guaranteed to generate a reset.

XTAL1: Input to the inverting Oscillator amplifier and input to the internal clock operating circuit.

XTAL2: Output from the inverting Oscillator amplifier.

AVCC: AVCC is the supply voltage pin for Port A and the A/D Converter. It should be externally connected to VCC, even if the ADC is not used. If the ADC is used, it should be con- nected to VCC through a low-pass filter.

AREF: AREF is the analog reference pin for the A/D Converter.

3.7.2 ATMEGA 8535 Architecture

The ATmega8535 provides the following features: 8K bytes of In-System Programmable Flash with Read-While-Write capabilities, 512 bytes EEPROM, 512 bytes SRAM, 32general purpose I/O lines, 32 general purpose working registers, three flexible Timer/Counters with compare modes, internal and external interrupts, a serial programmable USAR byte oriented Two wire Serial Interface, an 8-channel, 10-bit ADC with optional differential input stage with programmable gain in TQFP package ,programmable Watchdog Timer with Internal Oscillator, an SPI serial port, and six software selectable power saving modes. The Idle mode stops the CPU while allowing the SRAM, Timer/Counters, SPI port, and interrupt system to continue functioning. The Power-down mode saves the register contents but freezes the Oscillator, disabling all other chip functions until the next interrupt or Hardware Reset. In Power-save mode, the asynchronous timer continues to run, allowing the user to maintain a timer base while the rest of the device is sleeping. The ADC Noise Reduction mode stops the CPU and all I/O modules except asynchronous timer and ADC, to minimize switching noise during ADC conversions. In Standby mode, the crystal/resonator Oscillator is running while the rest of the device is sleeping. This allows very fast start-up combined with low-power consumption. In Extended Standby mode, both the main Oscillator and the asynchronous timer continue to run.

The ATmega8535 AVR is supported with a full suite of program and system development tools including: C compilers, macro assemblers, program debugger/simulators, In-Circuit Emulators, and evaluation kits.

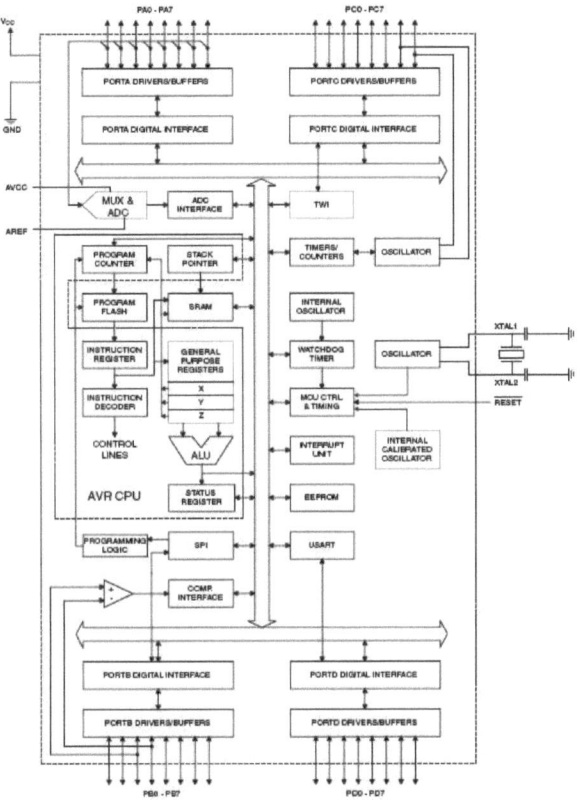

Fig 3.6 ATMEGA 8535 Architecture

3.7.3 Memory Organization

The ATmega8535 provides the following features: 8K bytes of In-System Programmable Flash with Read-While-Write capabilities, 512 bytes EEPROM, 512 bytes SRAM.

This section describes the different memories in the ATmega8535. The AVR architecture has two main memory spaces, the Data Memory and the Program Memory space. In addition, the ATmega8535 features an EEPROM Memory for data storage. All three memory spaces are linear and regular.

The ATmega8535 contains 8K bytes On-chip In-System Reprogrammable Flash memory for program storage. Since all AVR instructions are 16 or 32 bits wide, the Flash is organized as 4K x 16. For software security, the Flash Program memory space is divided into two sections, Boot Program section and Application Program section.

3.7.4 SRAM Data Memory

The 608 Data Memory locations address the Register File, the I/O Memory, and the internal data SRAM. The first 96 locations address the Register File and I/O Memory, and the next 512 locations address the internal data SRAM.

3.7.5 Special Function Registers (SFR)

Special function registers are part of RAM memory. Their purpose is predefined by the manufacturer and cannot be changed therefore. Since their bits are physically connected to particular circuits within the microcontroller, such as A/D converter, serial communication module etc., any change of their state directly affects the operation of the microcontroller or some of the circuits.

3.7.6 Program Counter

Program Counter is an engine running the program and points to the memory address containing the next instruction to execute. After each instruction execution, the value of the counter is incremented by 1. For this reason, the program executes only one instruction at a time just as it is written. However…the value of the program counter can be changed at any moment, which causes a "jump" to a new memory location.

3.7.7 Input/output ports (I/O Ports)

The ATMEGA 8535 has 32general purpose I/O lines, In order to make the microcontroller useful, it is necessary to connect it to peripheral devices. Each microcontroller has one or more registers (called a port) connected to the microcontroller pins. Why do we call them input/output ports? Because it is possible to change a pin function according to the user's needs. These registers are the only registers in the microcontroller the state of which can be checked by voltmeter.

3.7.8 Status Register

The Status Register contains information about the result of the most recently executed arithmetic instruction. This information can be used for altering program flow in order to perform conditional operations. Note that the Status Register is updated after all ALU operations, as specified in the Instruction Set Reference. This will, in many cases, remove the need for using the dedicated compare instructions, resulting in faster and more compact code The Status Register is not automatically stored when entering an interrupt routine and restored when returning from an interrupt. This must be handled by software.

Table 3.1. Status Register

Bit	7	6	5	4	3	2	1	0	
	I	T	H	S	V	N	Z	C	SREG
Read/Write	R/W	R/W	R/W	R/W	R/W	R/W	R/W	R/W	
Initial Value	0	0	0	0	0	0	0	0	

3.8 AT MEGA 8 MICROCONTROLLER

The Atmel® AVR® ATmega8 is a low-power CMOS 8-bit microcontroller based on the AVR RISC architecture. By executing powerful instructions in a single clock cycle, the ATmega8 achieves throughputs approaching 1MIPS per MHz, allowing the system designer to optimize power consumption versus processing speed. The Atmel® AVR® core combines a rich instruction set with 32 general purpose working registers. All the 32 registers are directly connected to the Arithmetic Logic Unit (ALU), allowing two independent registers to be accessed in one single instruction executed in one clock cycle. The resulting architecture is more code efficient while achieving throughputs up to ten times faster than conventional CISC microcontrollers.

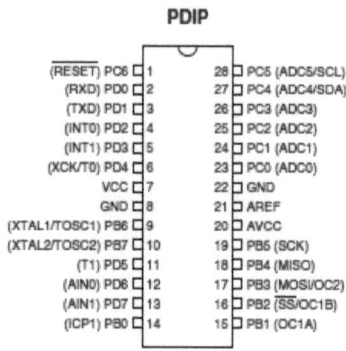

Fig.3.7 Pin Diagram of ATMEGA 8

3.8.1 Pin Description

VCC: Digital supply voltage.

GND: Ground.

Port B (PB7..PB0)XTAL1/XTAL2/TOSC1/TOSC2: Port B is an 8-bit bi-directional I/O port with internal pull-up resistors (selected for each bit). The Port B output buffers have symmetrical drive characteristics with both high sink and source capability. As inputs, Port B pins that are externally pulled low will source current if the pull-up resistors are activated. The Port B pins are tri-stated when a reset condition becomes active, even if the clock is not running. Depending on the clock selection fuse settings, PB6 can be used as input to the inverting Oscillator amplifier and input to the internal clock operating circuit. Depending on the clock selection fuse settings, PB7 can be used as output from the inverting Oscillator amplifier. If the Internal Calibrated RC Oscillator is used as chip clock source, PB7..6 is used as TOSC2..1input for the Asynchronous Timer/Counter2 if the AS2 bit in ASSR is set.

Port C (PC5..PC0): Port C is an 7-bit bi-directional I/O port with internal pull-up resistors (selected for each bit). The Port C output buffers have symmetrical drive characteristics with both high sink and source capability. As inputs, Port C pins that are externally pulled low will source current if the pull-up resistors are activated. The Port C pins are tri-stated when a reset condition becomes active, even if the clock is not running.

PC6/RESET: If the RSTDISBL Fuse is programmed, PC6 is used as an I/O pin. Note that the electrical characteristics of PC6 differ from those of the other pins of Port C. If the RSTDISBL Fuse is un programmed, PC6 is used as a Reset input. A low level on this pin for longer than the minimum pulse length will generate a Reset, even if the clock is not running generate a Reset.

Port D (PD7..PD0): Port D is an 8-bit bi-directional I/O port with internal pull-up resistors (selected for each bit). The Port D output buffers have symmetrical drive characteristics with both high sink and source capability. As inputs, Port D pins that are externally pulled low will source
current if the pull-up resistors are activated. The Port D pins are tri-stated when a reset condition becomes active, even if the clock is not running.

RESET: Reset input. A low level on this pin for longer than the minimum pulse length will generate areset, even if the clock is not running. Shorter pulses are not guaranteed to generate a reset.

3.8.2 ATMEGA 8 Architecture

The ATmega8 provides the following features: 8 Kbytes of In-System Programmable Flash with Read-While-Write capabilities, 512 bytes of EEPROM, 1 Kbyte of SRAM, 23 general purpose I/O lines, 32 general purpose working registers, three flexible Timer/Counters with compare modes, internal and external interrupts, a serial programmable USART, a byte oriented Two wire Serial Interface, a 6-channel ADC (eight channels in TQFP and QFN/MLF packages) with10-bit accuracy, a programmable Watchdog Timer with Internal Oscillator, an SPI serial port, and five software selectable power saving modes. The Idle mode stops the CPU while allowing the SRAM, Timer/Counters, SPI port, and interrupt system to continue functioning. The Power downmode saves the register contents but freezes the Oscillator, disabling all other chip functions until the next Interrupt or Hardware Reset. In Power-save mode, the asynchronous timer continues to run, allowing the user to maintain a timer base while the rest of the device is sleeping. The ADC Noise Reduction mode stops the CPU and all I/O modules except asynchronous timer and ADC, to minimize switching noise during ADC conversions. In Standby mode, the crystal/resonator Oscillator is running while the rest of the device is sleeping. This allows very fast start-up combined with low-power consumption.

The ATmega8 is supported with a full suite of program and system development tools, including C compilers, macro assemblers, program simulators, and evaluation kits.

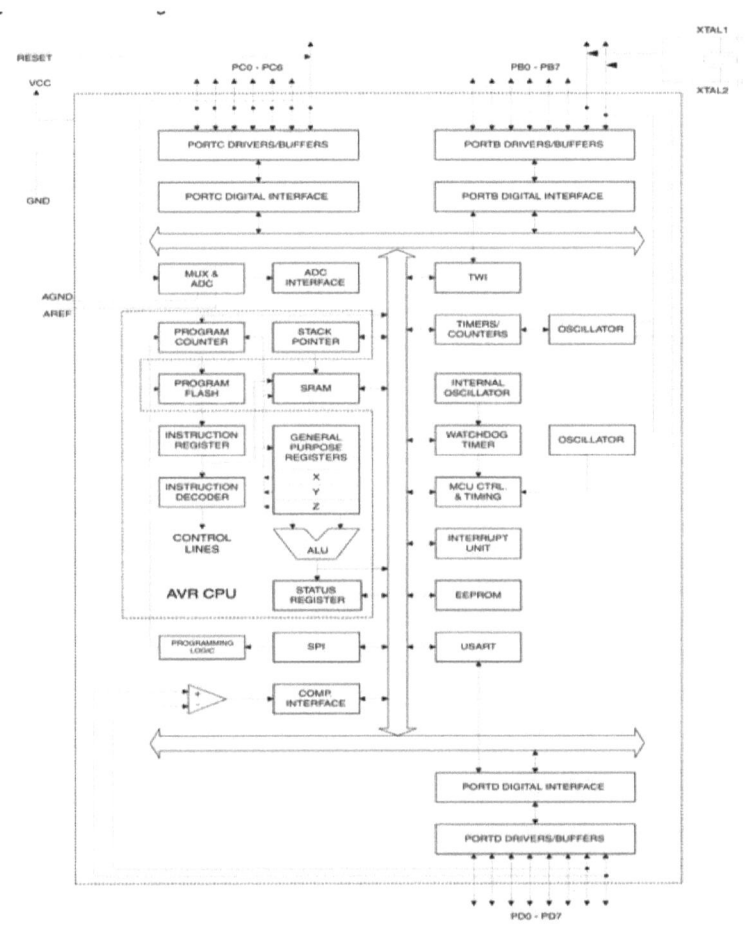

Fig 3.8 ATMEGA 8 Architecture

3.8.3 Memory Organization

The ATmega8 provides the following features: 8 Kbytes of In-System Programmable Flash with Read-While-Write capabilities, 512 bytes of EEPROM, 1 Kbyte of SRAM.

3.8.4 Special Function Registers (SFR)

Special function registers are part of RAM memory. Their purpose is predefined by the manufacturer and cannot be changed therefore. Since their bits are physically connected to particular circuits within the microcontroller, such as A/D converter, serial communication module etc., any change of their state directly affects the operation of the microcontroller or some of the circuits.

3.8.5 Program Counter

Program Counter is an engine running the program and points to the memory address containing the next instruction to execute. After each instruction execution, the value of the counter is incremented by 1. For this reason, the program executes only one instruction at a time just as it is written. However…the value of the program counter can be changed at any moment, which causes a "jump" to a new memory location.

3.8.6 Input/output ports (I/O Ports)

It has 23 Programmable I/O Lines, 28-lead PDIP, 32-lead TQFP, and 32-pad QFN/MLF and the Operating Voltages are 4.5V - 5.5V (ATmega8).

3.9 TARANG

Tarang 2.4GHz Modules are suitable for adding wireless capability to any product with serial data interface. The modules require minimal power and provide reliable delivery of data between devices. The I/O interfaces provided with the Module help to directly fit into many industrial applications.

The modules operate within the ISM 2.4 GHz frequency with 802.15.4 base band.

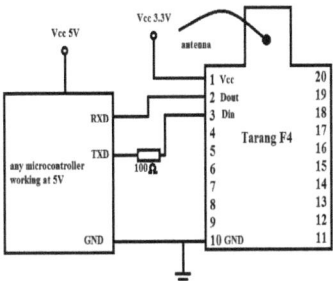

Fig.3.9 Tarang

A whip antenna is an antenna consisting of a single straight flexible wire or rod. The bottom end of the whip is connected to the radio receiver or transmitter. They are designed to be flexible so that they won't break off, and the name is derived from their whip-like motion when disturbed. Often whip antennas for portable radios are made of a series of interlocking telescoping metal tubes, so they can be retracted when not in use. Longer ones made for mounting on vehicles or structures are made of a flexible fiberglass rod surrounding a wire core, and can be up to 35 ft (10 m) long. Whips are the most common type of monopole antenna. These antennas are widely used for hand-held radios such as cell phones, cordless phones, walkie-talkies, FM radios, boom boxes, Wifi enabled devices, and GPS receivers, and also attached to vehicles as the antennas for car radios and two way radios for police, fire and aircraft. Larger versions mounted on roofs or radio masts are used as base station antennas for police, fire, ambulance, taxi and other vehicle dispatchers.

3.9.1 Features
- Point to point, point to multi point, Mesh and peer-to-peer topologies on proprietary stack.
- Direct Sequence Spread Spectrum technology.

- Each direct sequence channel has 64K unique network addresses.
- Transmit Power: 0 dBm.
- RF data rate: 250 kbps.
- Acknowledgement mode communication with retries.
- Power saving modes.
- Source / destination addressing.

3.10 LCD

Liquid Crystal Display (LCD) consists of rod-shaped tiny molecules sandwiched between a flat piece of glass and an opaque substrate. These rod-shaped molecules in between the plates align into two different physical positions based on the electric charge applied to them.

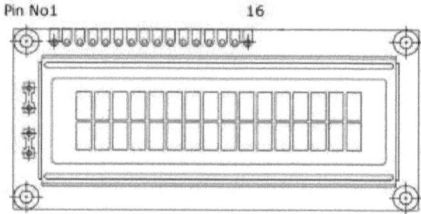

Fig.3.10 LCD

When electric charge is applied they align to block the light entering through them, where as when no-charge is applied they become transparent. Light passing through makes the desired images appear. This is the basic concept behind LCD displays. LCDs are most commonly used because of their advantages over other display technologies. They are thin and flat and consume very small amount of power compared to LED displays and cathode ray tubes (CRTs).

3.10.1 LCD Technologies and types

- Blue Mode STN - This is the basic LCD, which needs lot of improvement on contrast ratio and viewing angle.

- FSTN (Film STN) - Comes with an additional linearization film to offer better contrast.
- CSTN (color STN) - Layers of color filters are added to create up to 65,000 colors.
- DSTN (Double STN) - Improves contrast and eliminates any other colors appearing on the screen.

3.10.2 LCD characteristics

Liquid Crystals are very sensitive to constant electric fields. Only AC-voltages should be applied as DC voltages can cause an electrochemical reaction, which destroys the liquid crystals irreversibly. Temperature dependent and in a very cold or hot environment LCD may not work correctly. This is a reversible effect. Some displays need temperature compensation circuits to automatically adjust the applied LC voltage.

3.10.3 Advantages

- Consumes less power and generates less heat
- Saves lot of space compared picture tubes due to LCD's flatness
- Due to less weight and flatness LCDs are highly portable
- No flicker and fewer screens glare in LCDs to reduce eyestrain.

3.11 BUZZER

A buzzer or beeper is a signaling device, usually electronic, typically used in automobiles, household appliances such as a microwave oven, or game shows. It most commonly consists of a number of switches or sensors connected to a control unit that determines if and which button was pushed or a preset time has lapsed, and usually illuminates a light on the appropriate button or control panel, and sounds a warning in the form of a continuous or intermittent buzzing or beeping sound.

Initially this device was based on an electromechanical system which was identical to an electric bell without the metal gong (which makes the ringing noise).Often these units were anchored to a wall or ceiling and used the ceiling or wall as a sounding board.

3.11.1 Features

- Rated Frequency: 3,100Hz
- Operating Voltage: 3 - 20Vdc
- Current Consumption: 14mA @ 12Vd
- Dimensions: 22.5mm Diameter, 19mm High, 29mm between mounting holes.

3.12 IR SENSOR

Infrared (IR) light is electromagnetic radiation with a wavelength longer than that of visible light, measured from the nominal edge of visible red light at 0.7 micrometers, and extending conventionally to 300 micrometers.

These wavelengths correspond to a frequency range of approximately 430 to 1 THz, and include most of the thermal radiation emitted by objects near room temperature. Microscopically, IR light is typically emitted or absorbed by molecules when they change their rotational-vibrational movements.

3.12.1 Features

- High accuracy .
- High sensitivity (110 V/W) .
- Low resistance (50 ohm).
- Very good signal-to-noise-ratio .
- Good response time (40 ms) .
- Low cost thin film technology .

3.13 LED

A **light-emitting diode (LED)** is a two-lead semiconductor light source that resembles a basic pn-junction diode, except that an LED also emits light.When an LED's anode lead has a voltage that is more positive than its cathode lead by at least the LED's forward voltage drop, current flows. Electrons are able to recombine with holes within the device, releasing energy in the form of photons. This effect is called electroluminescence, and the color of the light (corresponding to the energy of the photon) is determined by the energy band gap of the semiconductor.

An LED is often small in area (less than1 mm^2), and integrated optical components may be used to shape its radiation pattern.

Appearing as practical electronic components in 1962, the earliest LEDs emitted low-intensity infrared light. Infrared LEDs are still frequently used as transmitting elements in remote-control circuits, such as those in remote controls for a wide variety of consumer electronics. The first visible-light LEDs were also of low intensity, and limited to red. Modern LEDs are available across the visible, ultraviolet, and infrared wavelengths, with very high brightness.

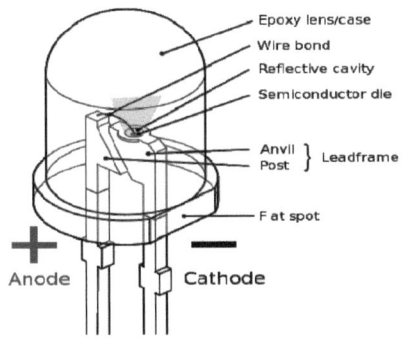

Fig.3.11. Structure Of LED

3.13.1 Advantages Of LED

- **Carbon emissions:** LEDs deliver significant reductions in carbon emissions
- **Color:** LEDs can emit light of an intended color without using any color filters as traditional lighting methods need. This is more efficient and can lower initial costs
- **Slow failure:** LEDs mostly fail by dimming over time, rather than the abrupt failure of incandescent bulbs.

3.14 POWER SUPPLY

Power supply is a reference to a source of electrical power. A device or system that supplies electrical or other types of energy to an output load or group of loads is called a power supply unit or PSU. The term is most commonly applied to electrical energy supplies, less often to mechanical ones, and rarely to others.

A 230v, 50Hz Single phase AC power supply is given to a step down transformer to get 12v supply. This voltage is converted to DC voltage using a Bridge Rectifier. The converted pulsating DC voltage is filtered by a 2200uf capacitor and then given to 7805 voltage regulator to obtain constant 5v supply. This 5v supply is given to all the components in the circuit. A RC time constant circuit is added to discharge all the capacitors quickly. To ensure the power supply a LED is connected for indication purpose.

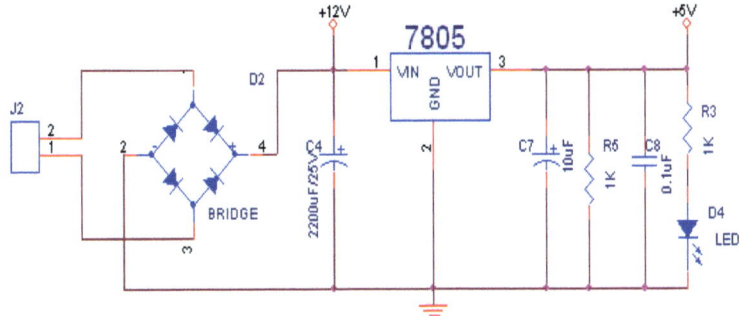

Fig 3.12. Circuit diagram of power supply

3.15 VOLTAGE REGULATOR: (IC 7805)

3.15.1 Features

- Output Current up to 1A
- Thermal Overload Protection
- Short Circuit Protection

3.15.2 Description

The KA78XX/KA78XXA series of three-terminal positive regulator are available in the TO-220/D-PAK package and with several fixed output voltages, making them useful in a wide range of applications. Each type employs internal current limiting, thermal shut down and safe operating area protection, making it essentially indestructible. If adequate heat sinking is provided, they can deliver over 1A output current.

3.16 RELAY

A **relay** is an electrically operated switch. Many relays use an electromagnet to mechanically operate a switch, but other operating principles are also used, such as solid-state relays. Relays are used where it is necessary to control a circuit by a low-power signal (with complete electrical isolation between control and controlled circuits), or where several circuits must be controlled by one signal.

The first relays were used in long distance telegraph circuits as amplifiers: they repeated the signal coming in from one circuit and re-transmitted it on another circuit. Relays were used extensively in telephone exchanges and early computers to perform logical operations.

A type of relay that can handle the high power required to directly control an electric motor or other loads is called a contactor. Solid-state relays control power circuits with no moving parts, instead using a semiconductor device to perform switching. Relays with calibrated operating characteristics and sometimes multiple operating coils are used to protect electrical circuits from overload or faults; in modern electric power systems these functions are performed by digital instruments still called "protective relays".

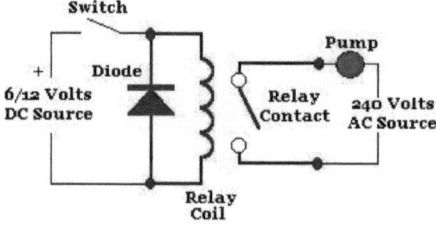

Fig.3.13 Relay

3.17 KEIL COMPILER

Keil Software is the leading vendor for 8/16-bit development tools (ranked at first position in the 2004 Embedded Market Study of the Embedded Systems and EE Times magazine). Keil Software is represented world-wide in more than 40 countries. Since the market introduction in 1988, the Keil C51 Compiler is the de facto industry standard and supports more than 500 current 8051 device variants. Now, Keil Software offers development tools for ARM.

Keil Software makes C compilers, macro assemblers, real-time kernels, debuggers, simulators, integrated environments, and evaluation boards for the 8051, 251, ARM, and XC16x/C16x/ST10 microcontroller families.

Keil Software is pleased to announce simulation support for the Atmel AT91 ARM family of microcontrollers. The Keil µVision Debugger simulates the complete ARM instruction-set as well as the on-chip peripherals for each device in the AT91 ARM/Thumb microcontroller family.

The integrated simulator provides complete peripheral simulation. Other new features in the µVision Debugger include:

- An integrated Software Logic Analyzer that measures I/O signals as well as program variables and helps developers create complex signal processing algorithms.
- An Execution Profiler that measures time spent in each function, source line, and assembler instruction. Now developers can find exactly where programs spend the most time.

Using nothing more than the provided simulation support and debug scripts, developers can create a high-fidelity simulation of their actual target hardware and environment. No extra hardware or test equipment is required.

The Logic Analyzer and Execution Profiler will help developers when it comes time to develop and tune signaling algorithms." said Jon Ward, President of Keil Software USA, Inc. Keil software is the leading vendor for 8/16-bit development tools (ranked at first position in the 2004 embedded market study of the embedded system and EE times magazine).

Keil software is represented worldwide in more than 40 countries, since the market introduction in 1988; the keil C51 compiler is the de facto industry standard and supports more than 500 current 8051 device variants. Now, keil software offers development tools for ARM.

Keil software makes C compilers, macro assemblers, real-time kernels, debuggers, simulators, integrated environments, and evaluation boards for 8051, 251, ARM and XC16x/C16x/ST10 microcontroller families.

The Keil C51 C Compiler for the 8051 microcontroller is the most popular 8051 C compiler in the world. It provides more features than any other 8051 C compiler available today.The C51 Compiler allows you to write 8051 microcontroller applications in C that, once compiled, have the efficiency and speed of assembly language. Language extensions in the C51 Compiler give you full access to all resources of the 8051.

The C51 Compiler translates C source files into relocatable object modules which contain full symbolic information for debugging with the µVision Debugger or an in-circuit emulator. In addition to the object file, the compiler generates a listing file which may optionally include symbol table and cross reference

Nine basic data types, including 32-bit IEEE floating-point,
- Flexible variable allocation with bit, data, bdata, idata, xdata, and pdata memory types,
- Interrupt functions may be written in C,
- Full use of the 8051 registers banks,
- Complete symbol and type information for source-level debugging,
- Use of AJMP and ACALL instructions,
- Bit-addressable data objects,
- Built-in interface for the RTX51 real time kernels,
- Support for the Philips 8xC750, 8xC751, and 8xC752 limited instruction sets,
- Support for the Infineon 80C517 arithmetic unit.

3.18 EMBEDDED C

Most common programming languages for embedded systems are C BASIC and assembly languages. C used for embedded systems is slightly different compared to C used for general purpose (under a PC platform). Programs for embedded systems are usually expected to monitor and control external devices and directly manipulate and use the internal architecture of the processor such as interrupt handling, timers, serial communications and other available features.- there are many factors to consider when selecting languages for embedded systems

- Efficiency - Programs must be as short as possible and memory must be used efficiently.
- Speed - Programs must run as fast as possible.
- Ease of implementation
- Maintainability
- Readability

C compilers for embedded systems must provide ways to examine and utilize various features of the microcontroller's internal and external architecture.

Most embedded C compilers (as well as ordinary C compilers) have been developed supporting the ANSI but compared to ordinary C they may differ in terms of the outcome of some of the statements. Standard C compiler, communicates with the hardware components via the operating system of the machine but the C compiler for the embedded system must communicate directly with the processor and its components.

3.19 CONCLUSION

Thus the Proposed system comprising of various components used in the Project and are explained in detail.

CHAPTER 4

RESULTS AND DISCUSSION

4.1 INTRODUCTION

The Project's output snapshots are discussed in this chapter,(i.e)The Transmitter and the Receiver Unit.

4.2 CAR UNIT

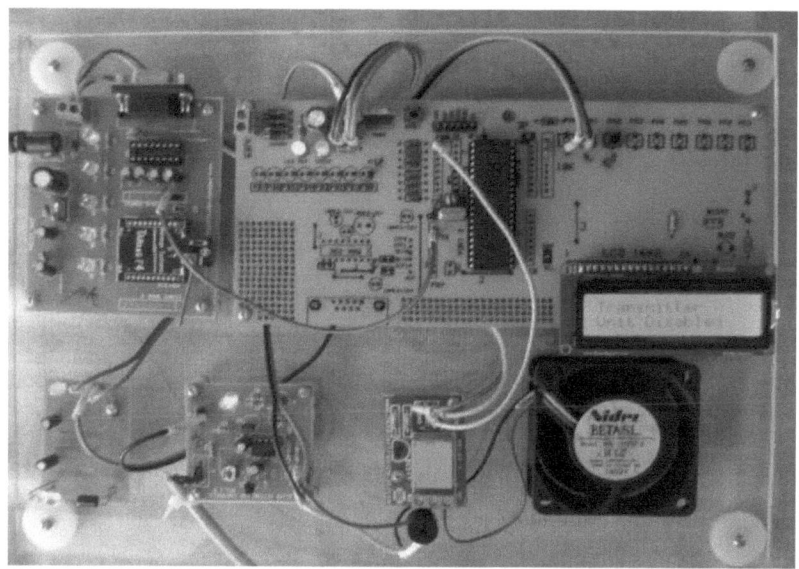

Fig.4.1 Image of Car Unit

4.2.1 Inference

The Fig 4.1 shows the circuit for Car Unit. It consists of Eye Blinking IR Sensor, spo2 sensor, Relay, controller (ATMEGA 8535), Motor, LCD, Buzzer.

4.2.2 Eye Blinking IR Sensor

IR Transmitter: Fig.4.2 Shows The IR Transmitter. A thermographic camera or infrared camera is a device that forms an image using infrared radiation, similar to a common camera that forms an image using visible light. Instead of the 450–750 nanometer range of the visible light camera, infrared cameras operate in wavelengths as long as 14,000 nm (14 µm).

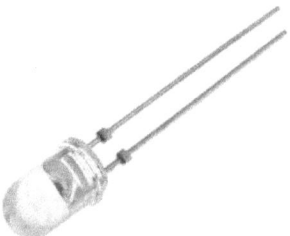

Fig.4.2 IR Transmitter

IR Receiver: Fig 4.3 shows the IR receiver. The IR beam data is received by an IrDA device equipped with a silicon photodiode. This receiver converts the IR beam into an electric current for processing. Because IR transitions more slowly from ambient light than from a rapidly pulsating IrDA signal, the silicon photodiode can filter out the IrDA signal from ambient IR. IrDA transmitters and receivers are classified as directed and non-directed.

Fig.4.3 IR Receiver

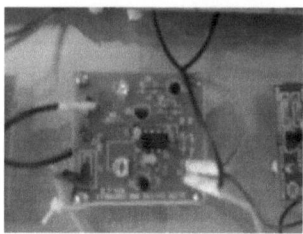

Fig.4.4 LM 358 IC Connected With Eye Blinking Sensor Circuit

LM358: Fig.4.5 shows the LM358 Op-Amp. It consists of two independent, high gain, internally frequency compensated operational amplifiers which were designed specifically to operate from a single power supply over a wide range of voltages. Operation from split power supplies is also possible and the low power supply current drain is independent of the magnitude of the power supply voltage.

Fig 4.5 A LM358 Dual Op-Amp

4.2.3 Heart Beat IR Sensor

Pulse Detector is a circuit which is used to measure the pulse at the finger tips.

Fig.4.6 Shows The Pulse Detection Sensor. A person's pulse can be detected by monitoring blood pressure changes within the human body. As the heart muscle contracts, blood pressure goes up and as it relaxes, blood pressure goes down. One clever method of monitoring blood pressure is to shine light through an appendage of the body and to measure the amount of light reaching a sensor.

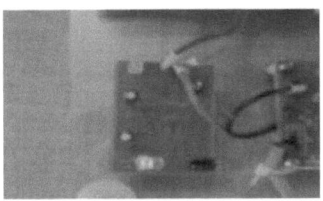

Fig.4.6 Pulse Detection sensor

4.2.4 TarangF4

The Fig.4.7 Shows the TarangF4 module. Modules are suitable for adding wireless capability to any product with serial data interface. The modules require minimal power and provide reliable delivery of data between devices. The I/O interfaces provided with the Module help to directly fit into many industrial applications. The modules operate within the ISM 2.4 GHz frequency with 802.15.4 base band.

Fig.4.7 Tarang Module

4.2.5 Working

The Car Unit is mounted near the driver's seat. If the driver is suffered by any attack or he is sleeping the controller warns the patient by a buzzer and the pulse rate, eye blinking time is displayed on the LCD screen. If the patient is normal only the pulse rate, eye blinking time is displayed on the LCD screen. If the patient is sleeping or pulse rate downs due to heart attack is indicated by a buzzer and the patient's record is transmitted to the nearby hospital by the Tarang F4. Then the motor is Turned OFF.

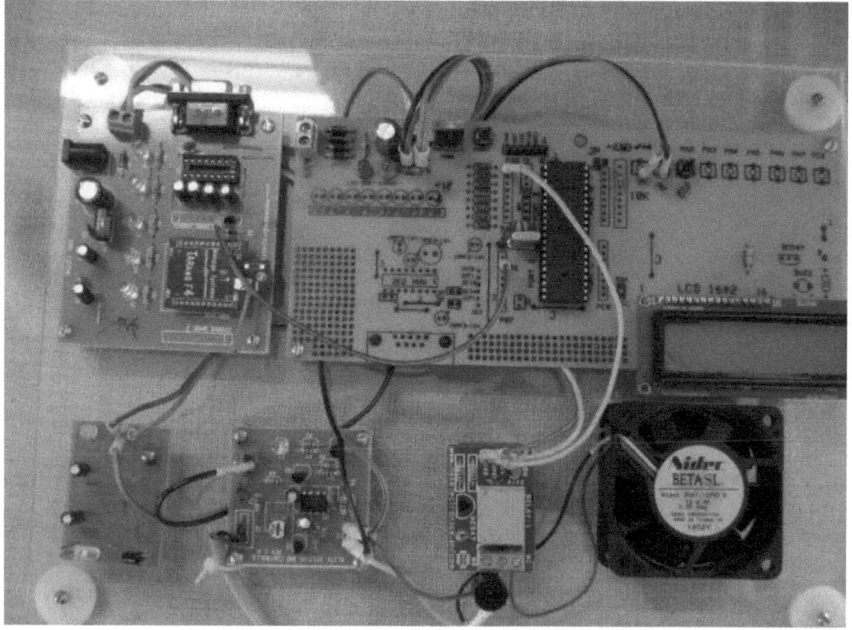

Fig.4.8 Top view of Transmitter car unit

The Fig.4.8 shows the Top view of the Transmitter Car Unit consists of Eye and Pulse detecting sensor,TarangF4,Controller,LCD and Buzzer.

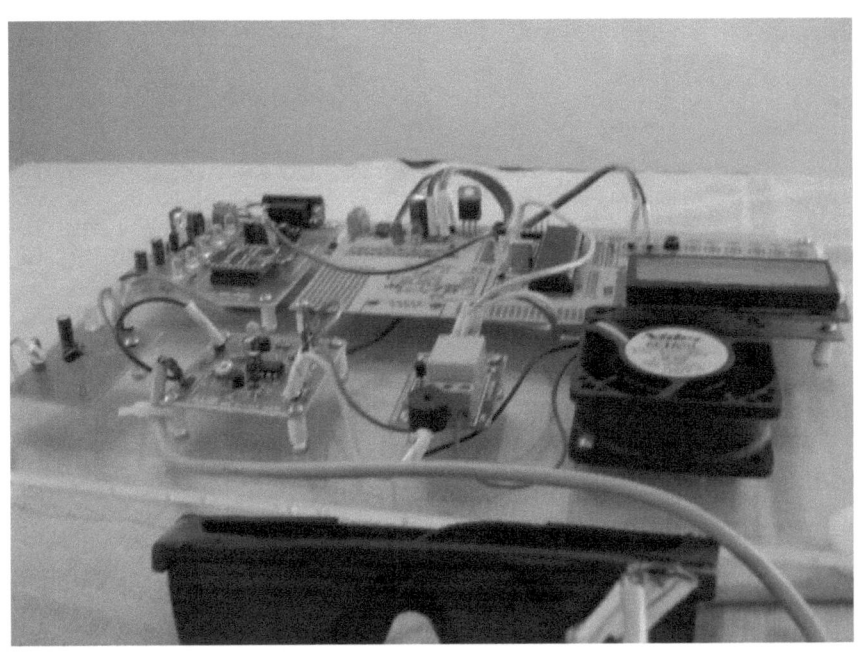

Fig.4.9 Front view of Transmitter car unit

The Fig.4.9 shows the Front View of Transmitter car unit.

4.3 HOSPITAL UNIT

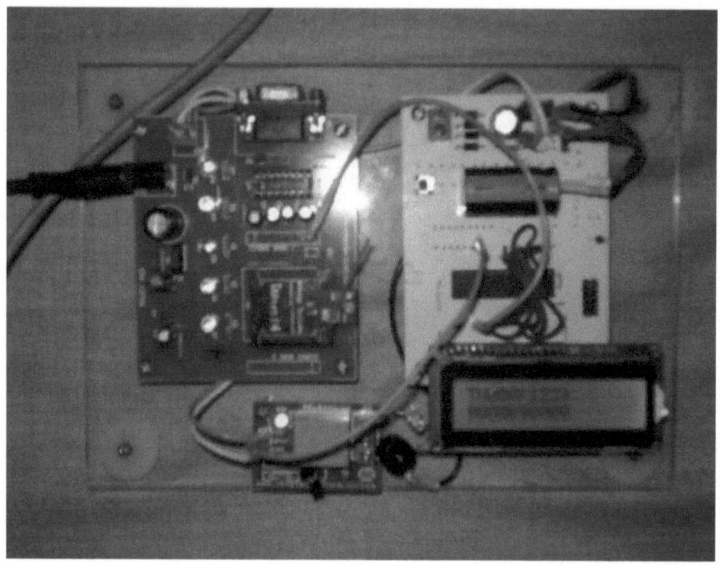

Fig.4.10 Image Of Hospital Unit

4.3.1 Inference

Fig.4.10 shows the circuit of Hospital Unit. It consists of Tarang, controller(ATMEGA 8), LCD, LED And Buzzer.

4.3.2 Buzzer

Fig.4.11 shows the Buzzer indicates the hospital about the patients by an Alarming sound.

Piezo buzzer is an electronic device commonly used to produce sound. Light weight, simple construction and low price make it usable in various applications like car/truck reversing indicator, computers, call bells.

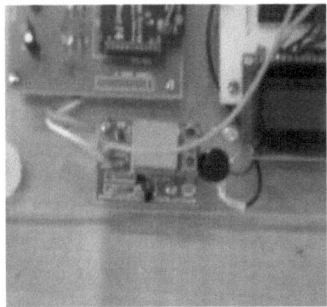

Fig.4.11 Buzzer

4.3.3 LCD Display

Fig.4.12 shows a 16x2 LCD which means it can display 16 characters per line and there are 2 such lines.

Fig.4.12 LCD

4.2.4 Working

In Case of any critical situations, the car engine is turned off by using solenoid valve to drain the fuel. The patient's information is sent to the nearby hospital having Tarang module. The patient's car number & mobile number is displayed on the LCD Screen of the Hospital Unit. and is indicated with a Buzzer.

CHAPTER 5

CONCLUSION AND FUTURE ENHANCEMENT

5.1 CONCLUSION

It was very satisfying to implement our own ideas and see it to fruition. We met the specifications that we set, and achieved our goal. This project presents a cheap, portable system for Heart Patients that is easily adapted to existing equipment. The Proposed method has the patient's information which is only displayed. If any delay or pulse down then the communication between the car and the hospital begins. In addition, the capabilities for identifying movement type are examined for the IR sensors.

The system is designed to be usable by all active personnel ,Since our hardware is intended to be mounted on existing equipment, there is no special requirement for using it. Only the communication between the hospital and vehicle is done &the vehicle and mobile number is displayed on the hospital unit.

5.2 SCOPE FOR THE FUTURE ENHANCEMENT

The wearable dress that consists of components in miniature form. The ECG device is added for more accurate results. The Wireless based sensors are being used. The IR Sensors are replaced, because it's slightly injurious. The Communication can be Worldwide.

5.3 APPLICATIONS

- Useful for Heart patients.
- Health Care Centres.
- Used in Vehicles for Accident avoidance.
- Mobile patient's Heart Monitoring System.

REFERENCES

[1] Myung-kyung Suh; Evangelista, L.S.; Chen, V.; Wen-Sao Hong; Macbeth, J.; Nahapetian, A.; Figueras, F.; Sarrafzadeh, M (2010), 'WANDA B.: Weight and activity with blood pressure monitoring system for heart failure patients', IEEE International Symposium on a , vol., no., pp.1,6, 14-17 June 2010

[2] Ren-Guey Lee; Yih-Chien Chen; Chun-Chieh Hsiao; Chwan-Lu Tseng,(2007), 'A Mobile Care System With Alert Mechanism,' Information Technology in Biomedicine, IEEE Transactions on , vol.11, no.5, pp.507,517, Sept. 2007

[3] Mahananto, F.; Igasaki, T.; Murayama,(2013), N., 'Cardiac arrhythmia detection using combination of heart rate variability analyses and PUCK analysis,' Engineering in Medicine and Biology Society (EMBC), 2013 35th Annual International Conference of the IEEE , vol., no., pp.1696,1699, 3-7 July 2013

[4] Jain, N.P.; Jain, P.N.; Agarkar, T.P.,(2012), "An embedded, GSM based, multiparameter, realtime patient monitoring system and control — An implementation for ICU patients," Information and Communication Technologies (WICT), 2012 World Congress on , vol., no., pp.987,992, Oct. 30 2012-Nov. 2 2012

[5] Wei Lin,(2011), "Real time monitoring of electrocardiogram through IEEE802.15.4 network," Emerging Technologies for a Smarter World (CEWIT), 2011 8th International Conference & Expo on , vol., no., pp.1,6, 2-3 Nov. 2011

[6] Palantei, E.; Baharuddin, M.; Andani, A.; Nien, K.N.; Utami, D.; Febriani, A. E A; Umar, U.; Agus, M.,(2012), "A 2.5 GHz wireless ECG system for remotely monitoring heart pulses," Antennas and Propagation Society International Symposium (APSURSI), 2012 IEEE , vol., no., pp.1,2, 8-14 July 2012

[7] Apostu, O.; Hagiu, B.; Pasca, S.,(2011), "Wireless ECG monitoring and alarm system using ZigBee," Advanced Topics in Electrical Engineering (ATEE), 2011 7th International Symposium on , vol., no., pp.1,4, 12-14 May 2011

[8] Yongwon Jang; Hyung Wook Noh; Lee, I. B.; Ji-Wook Jung; Yoonseon Song; Sooyeul Lee; Seunghwan Kim,(2012), "Development of a patch type embedded cardiac function monitoring system using dual microprocessor for arrhythmia detection in heart disease patient," Engineering in Medicine and Biology Society (EMBC), 2012 Annual International Conference of the IEEE , vol., no., pp.2162,2165, Aug. 28 2012-Sept. 1 2012

[9] Rotariu, C.; Pasarica, A.; Costin, H.; Adochiei, F.; Ciobotariu, R., (2011),"Telemedicine system for remote blood pressure and heart rate monitoring," E-Health and Bioengineering Conference (EHB), 2011 , vol., no., pp.1,4, 24-26 Nov. 2011

[10] De Capua, C.; Meduri, A.; Morello, R.,(2010), "A Smart ECG Measurement System Based on Web-Service-Oriented Architecture for Telemedicine Applications," Instrumentation and Measurement, IEEE Transactions on , vol.59, no.10, pp.2530,2538, Oct. 2010

APPENDIX

PROGRAM FOR ATMEGA 8535-TRANSMITTER CAR UNIT

```c
#include <mega8535.h>
#include<delay.h>

// Alphanumeric LCD Module functions
#asm
   .equ __lcd_port=0x15 ;PORTC
#endasm
#include <lcd.h>

// Standard Input/Output functions
#include <stdio.h>
int eyeblink=0;
int blink=0;
int time_count=0;
int sec=0;
int bpm=1;
int beat=0;
int disable=1;
char disp[20];
// Timer 0 overflow interrupt service routine
interrupt [TIM0_OVF] void timer0_ovf_isr(void)
{
// Place your code here
TCNT0=0;
if(++time_count==100)
{
sec++;
time_count=0;
```

```
}
if(sec>60)
sec=0;
if(PINA.0==1)
{
beat++;
while(PINA.0==1);
}
if(sec==15)
{
bpm=beat*4;
sprintf(disp,"BPM=%d",bpm);
lcd_gotoxy(0,1);
lcd_putsf("            ");
lcd_gotoxy(0,1);
lcd_puts(disp);
beat=0;
sec=0;
}
if(PINA.1==0)
{
blink++;
}
if(blink==400)
{
blink=0;
eyeblink=1;
lcd_gotoxy(0,1);
lcd_putsf("            ");
```

```
lcd_gotoxy(0,1);
lcd_putsf("Eye closed");
}
if(PINA.1==1)
{
blink=0;
}
}
```

// Declare your global variables here

```
void main(void)
{
// Declare your local variables here



// Input/Output Ports initialization
// Port A initialization
// Func7=In Func6=In Func5=In Func4=In Func3=In Func2=In Func1=In Func0=In
// State7=T State6=T State5=T State4=T State3=T State2=T State1=T State0=T
PORTA=0x00;
DDRA=0x00;

// Port B initialization
// Func7=Out Func6=Out Func5=Out Func4=Out Func3=Out Func2=Out Func1=Out Func0=Out
// State7=0 State6=0 State5=0 State4=0 State3=0 State2=0 State1=0 State0=0
PORTB=0x00;
```

DDRB=0xFF;

// Port C initialization
// Func7=In Func6=In Func5=In Func4=In Func3=In Func2=In Func1=In Func0=In
// State7=T State6=T State5=T State4=T State3=T State2=T State1=T State0=T
PORTC=0x00;
DDRC=0x00;

// Port D initialization
// Func7=In Func6=In Func5=In Func4=In Func3=In Func2=In Func1=In Func0=In
// State7=T State6=T State5=T State4=T State3=T State2=T State1=T State0=T
PORTD=0x00;
DDRD=0x00;

// Timer/Counter 0 initialization
// Clock source: System Clock
// Clock value: 31.250 kHz
// Mode: Normal top=FFh
// OC0 output: Disconnected
TCCR0=0x04;
TCNT0=0x00;
OCR0=0x00;

// Timer/Counter 1 initialization
// Clock source: System Clock
// Clock value: Timer 1 Stopped
// Mode: Normal top=FFFFh

```
// OC1A output: Discon.
// OC1B output: Discon.
// Noise Canceler: Off
// Input Capture on Falling Edge
// Timer 1 Overflow Interrupt: Off
// Input Capture Interrupt: Off
// Compare A Match Interrupt: Off
// Compare B Match Interrupt: Off
TCCR1A=0x00;
TCCR1B=0x00;
TCNT1H=0x00;
TCNT1L=0x00;
ICR1H=0x00;
ICR1L=0x00;
OCR1AH=0x00;
OCR1AL=0x00;
OCR1BH=0x00;
OCR1BL=0x00;

// Timer/Counter 2 initialization
// Clock source: System Clock
// Clock value: Timer 2 Stopped
// Mode: Normal top=FFh
// OC2 output: Disconnected
ASSR=0x00;
TCCR2=0x00;
TCNT2=0x00;
OCR2=0x00;
```

// External Interrupt(s) initialization
// INT0: Off
// INT1: Off
// INT2: Off
MCUCR=0x00;
MCUCSR=0x00;

// Timer(s)/Counter(s) Interrupt(s) initialization
TIMSK=0x01;

// USART initialization
// Communication Parameters: 8 Data, 1 Stop, No Parity
// USART Receiver: Off
// USART Transmitter: On
// USART Mode: Asynchronous
// USART Baud rate: 9600
UCSRA=0x00;
UCSRB=0x08;
UCSRC=0x00;
UBRRH=0x00;
UBRRL=0x00;

// Analog Comparator initialization
// Analog Comparator: Off
// Analog Comparator Input Capture by Timer/Counter 1: Off
ACSR=0x80;
SFIOR=0x00;

// LCD module initialization

```c
lcd_init(16);
lcd_putsf("Transmitter");
delay_ms(2000);

// Global enable interrupts
//#asm("sei")

while (1)
    {
    // Place your code here
    if(bpm==0||eyeblink==1)
    {
    PORTB.0=1;
    putchar(0xAA);
    delay_ms(300);
    putsf("TN00AK1234");
    //putchar(0xAB);
    //putsf("9000090000");
    delay_ms(2000);

    }


    if(disable==1)
    {
    #asm("cli")
    lcd_gotoxy(0,1);
    lcd_putsf("            ");
    lcd_gotoxy(0,1);
```

```
    lcd_putsf("Unit Disabled");
    while(disable==1)
    {
     if(PINA.2==1)
     {
     #asm("sei")
     lcd_gotoxy(0,1);
     lcd_putsf("            ");
     lcd_gotoxy(0,1);
     lcd_putsf("Unit Enabled");
     disable=0;
     while(PINA.2==1);
     delay_ms(500);
     }
    }
    }

    if(PINA.2==1)
    {
    if(disable==1)
    disable=0;
    else
    disable=1;
    while(PINA.2==1);
    delay_ms(500);
    }
    };
}
```

PROGRAM FOR ATMEGA 8-RECEIVER HOSPITAL UNIT

```c
#include <mega8.h>
#include <delay.h>
// Alphanumeric LCD Module functions
#asm
   .equ __lcd_port=0x18 ;PORTB
#endasm
#include <lcd.h>

// Standard Input/Output functions
#include <stdio.h>
unsigned char rec[10];

// Declare your global variables here

void main(void)
{
// Declare your local variables here

// Input/Output Ports initialization
// Port B initialization
// Func7=In Func6=In Func5=In Func4=In Func3=In Func2=In Func1=In Func0=In
// State7=T State6=T State5=T State4=T State3=T State2=T State1=T State0=T
PORTB=0x00;
DDRB=0x00;

// Port C initialization
```

// Func6=Out Func5=Out Func4=Out Func3=Out Func2=Out Func1=Out Func0=Out
// State6=0 State5=0 State4=0 State3=0 State2=0 State1=0 State0=0
PORTC=0x00;
DDRC=0x7F;

// Port D initialization
// Func7=In Func6=In Func5=In Func4=In Func3=In Func2=In Func1=In Func0=In
// State7=T State6=T State5=T State4=T State3=T State2=T State1=T State0=T
PORTD=0x00;
DDRD=0x00;

// Timer/Counter 0 initialization
// Clock source: System Clock
// Clock value: Timer 0 Stopped
TCCR0=0x00;
TCNT0=0x00;

// Timer/Counter 1 initialization
// Clock source: System Clock
// Clock value: Timer 1 Stopped
// Mode: Normal top=FF
// OC1A output: Discon.
// OC1B output: Discon.
// Noise Canceler: Off
// Input Capture on Falling Edge
// Timer 1 Overflow Interrupt: Off
// Input Capture Interrupt: Off

// Compare A Match Interrupt: Off
// Compare B Match Interrupt: Off
TCCR1A=0x00;
TCCR1B=0x00;
TCNT1H=0x00;
TCNT1L=0x00;
ICR1H=0x00;
ICR1L=0x00;
OCR1AH=0x00;
OCR1AL=0x00;
OCR1BH=0x00;
OCR1BL=0x00;

// Timer/Counter 2 initialization
// Clock source: System Clock
// Clock value: Timer 2 Stopped
// Mode: Normal top=F
// OC2 output: Disconnected
ASSR=0x00;
TCCR2=0x00;
TCNT2=0x00;
OCR2=0x00;

// External Interrupt(s) initialization
// INT0: Off
// INT1: Off
MCUCR=0x00;

// Timer(s)/Counter(s) Interrupt(s) initialization

```
TIMSK=0x00;

// USART initialization
// Communication Parameters: 8 Data, 1 Stop, No Parity
// USART Receiver: On
// USART Transmitter: Off
// USART Mode: Asynchronous
// USART Baud rate: 9600
UCSRA=0x00;
UCSRB=0x10;
UCSRC=0x86;
UBRRH=0x00;
UBRRL=0x33;

// Analog Comparator initialization
// Analog Comparator: Off
// Analog Comparator Input Capture by Timer/Counter 1: Off
ACSR=0x80;
SFIOR=0x00;

// LCD module initialization
lcd_init(16);
lcd_putsf("Receiver");
while (1)
    {
    // Place your code here
    if(getchar()==0xAA)
    {
    PORTC.0=1;
```

```
        gets(rec,10);
        lcd_clear();
        lcd_puts(rec);
        lcd_gotoxy(0,1);
        lcd_putsf("9000090000");


     }
    };
}
```